This book is dedicated to all the underdogs who have yet to get the recognition they deserve.

Garrett's Store
A Tiny Thinkers Book

Written by M.J. Mouton
Illustrated by: Jezreel Cuevas
Edited by: Amanda N. Franquet

Art Consultant: Deanie Mouton
Book Design: Amanda N. Franquet

US Library of Congress
ISBN: 978-09983147-4-7

Hi, I'm Hitch!

I've spent time with some amazing Tiny Thinkers! Join me as we learn about people and the science they discovered. And see if you can spot me along the way, as I tell you the story of Garrett's real-life adventure that changed the world!

Garrett's Store

Written by M.J. Mouton Illustrated by Jezreel S. Cuevas

A Foreword by Dr. Kristina Henry Collins

I am a professor and social scientist. I study how people teach, learn, and treat other people. Before that, I was a mathematician and before that I was an engineer. Like Garrett, I enjoy exploring how things work to make them work better or to invent new things so that we can work better. And just like Garrett, I often times wondered why would people treat scientists, engineers, and inventors different if they are good at what they do.

Garrett invented and improved many things. He did so because he wanted to make life and work better for everyone. But there were some people that believed that only inventors that looked like them – had the same color of hair, eyes, or even face – could invent things that work. This was not true. And some people just didn't want to buy from Garrett because he looked different from them. This was not fair either. So, Garrett hired people that looked more like the customers that would come into his store to pretend like they invented Garrett's things. He would pretend to be an assistant sometimes, too.

Garrett had to work much harder to be successful and to earn people's trust because they did not know him. They did not take the time to get to know him. You may think it was wrong that he "tricked" people into buying things from his store, but what he was trying to do was prove to people that he was a good inventor too, and they would know that if they only gave him a chance. In this story, the customers may not have known Big Chief Mason or the clerk either, but because Big Chief Mason or the clerk looked like them in some ways, they automatically trusted them, and bought from their store. In my work, we call this "privilege".

I learned about Garrett when I was in school. His story inspired me to become an inventor. When I was studying engineering, I invented a "Reflex Tester" as a project for one of my classes. I was very proud of it. And when people did not trust me because I was a girl, I remembered how Garrett did not let that stop him from inventing and improving things to help people work better. That's what I wanted to do. Now, I help students that look like me who want to be scientists, technicians, engineers, and mathematicians – also known as STEM professionals. I help them to understand how important it is that they follow their dreams, and help them to believe in themselves even when others may not.

As you read this story about "Garrett's Store", I hope that you are inspired to create and invent things to make people live and work better. Everyone – no matter how different we are from each other – can help make the world a better place.

Dr. Kristina Henry Collins
Educational & Research Psychologist
STEM Identity & Talent Development

Chief

On a very cold morning, Garrett opened his store.
He hoped today would be better than the day before.

People rarely purchased any of
their needs from Garrett.
Garrett did not understand
why this was.
He hoped it was not what
seemed too apparent.
“Does my breath smell bad?
Is my hair in a mess?
Am I smiling enough?
Am I properly dressed?”

Garrett asked, "Why will they not buy a thing from my store?"
As he turned over the bright OPEN sign on his door.

Garrett's store was by far the cheapest store around,
But many people would not shop at his store in this town.

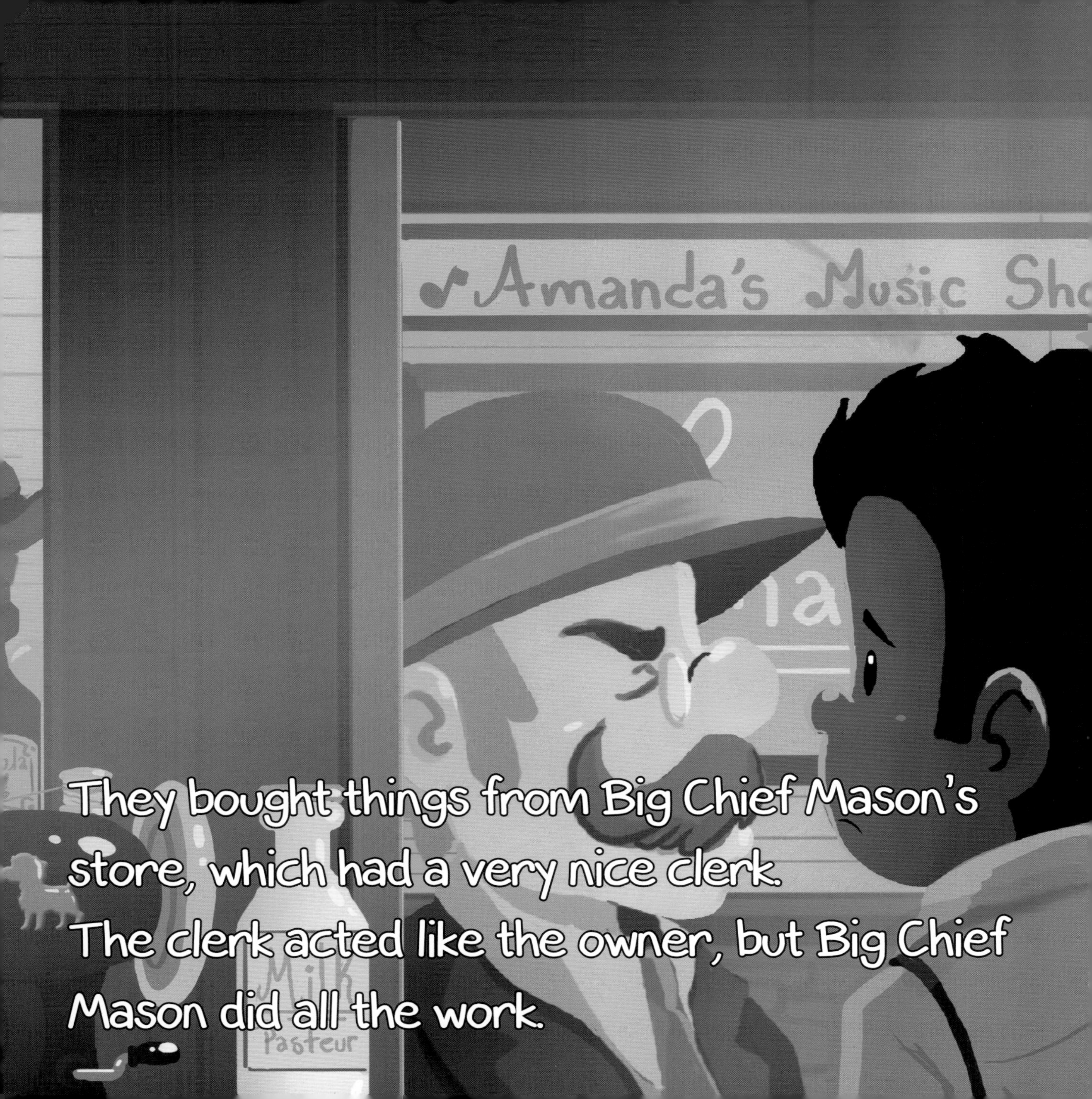

They bought things from Big Chief Mason's store, which had a very nice clerk.
The clerk acted like the owner, but Big Chief Mason did all the work.

Garrett would always say, "If you will not buy from me, then shop at Big Chief Mason's. He is on this very same street.

I will say this just once, I will not repeat:
I have every item that Chief Mason has at his store,
But every item in his place will cost one nickel more."

Garrett wondered, "Do they buy from Big Chief Mason because he has fantastic teeth?
Is it because you can see your reflection in the shoes upon his feet?
Is it because they like the feathers that he wears upon his head,
Or do they like his war paint, white and red?"

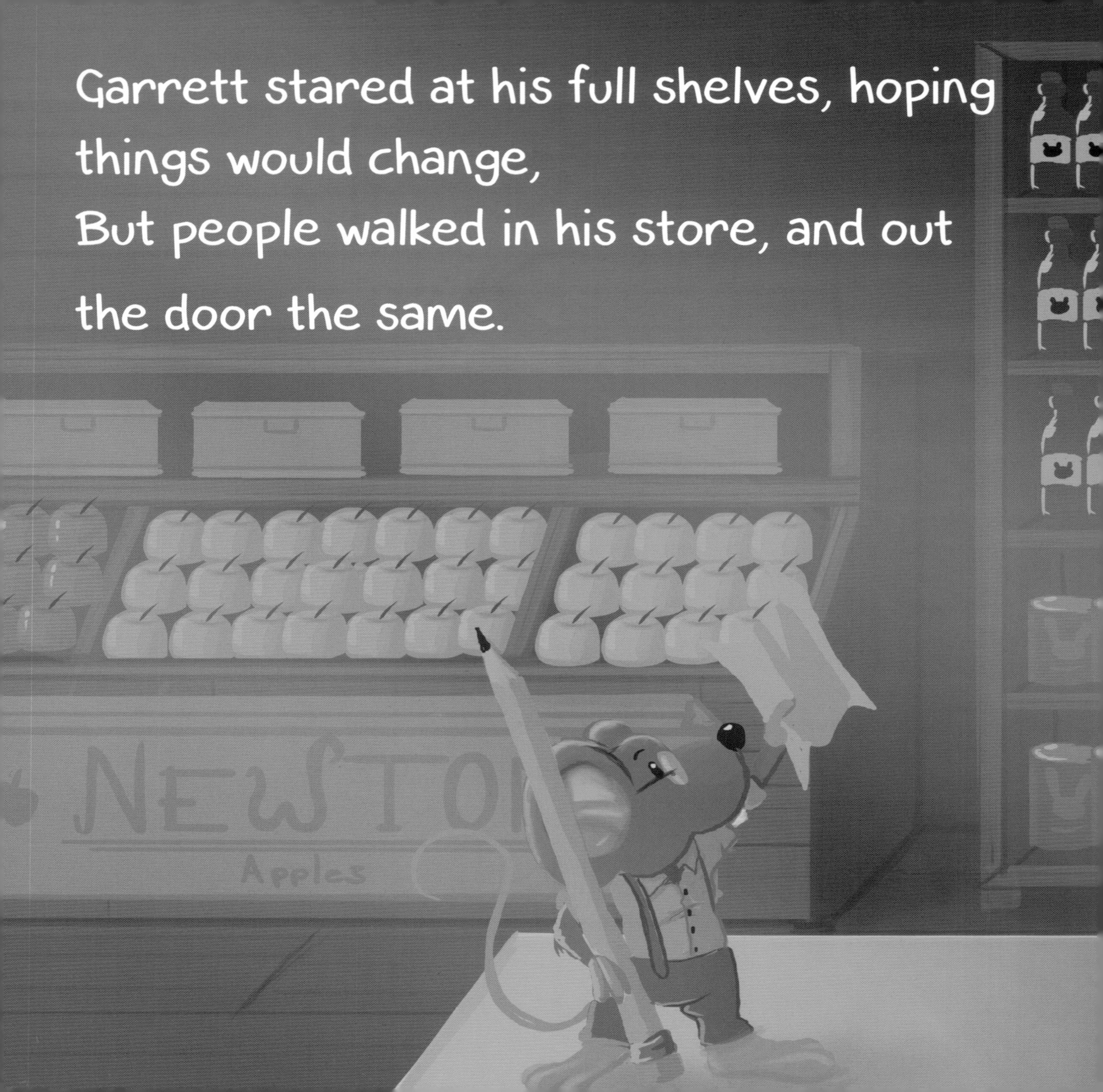

Garrett stared at his full shelves, hoping
things would change,
But people walked in his store, and out
the door the same.

They walked right out of Garrett's place and he almost knew for sure,
That they would walk to Mason's, and pay one nickel more.

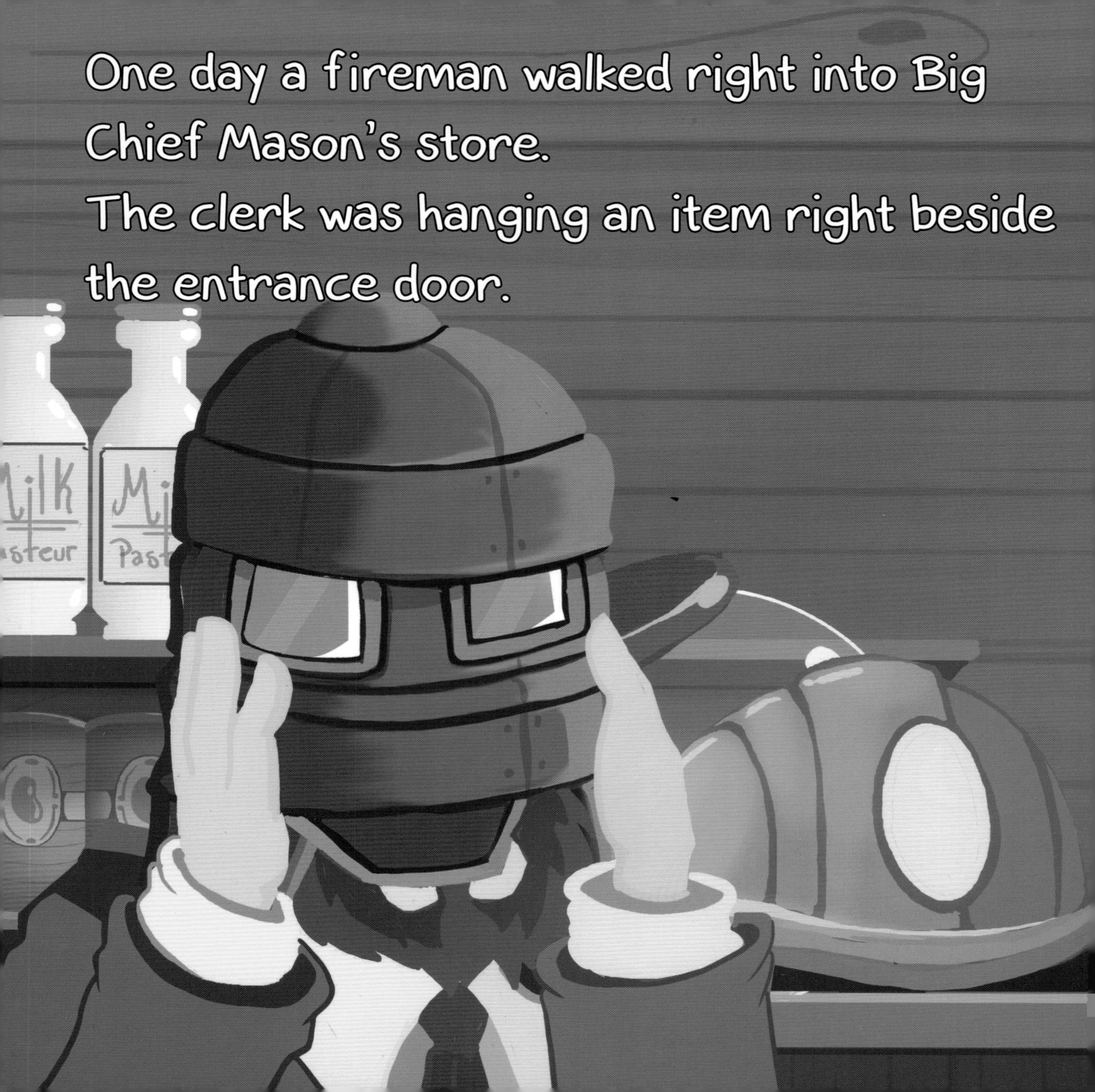

One day a fireman walked right into Big Chief Mason's store.
The clerk was hanging an item right beside the entrance door.

'What is this?" asked
:he fireman. "It
ooks like a mask."

To which the clerk replied, "I'm glad that you asked. It is a device that helps you walk into a building that's on fire.
It helps you breathe air from near the floor because the smoke will stay up higher."

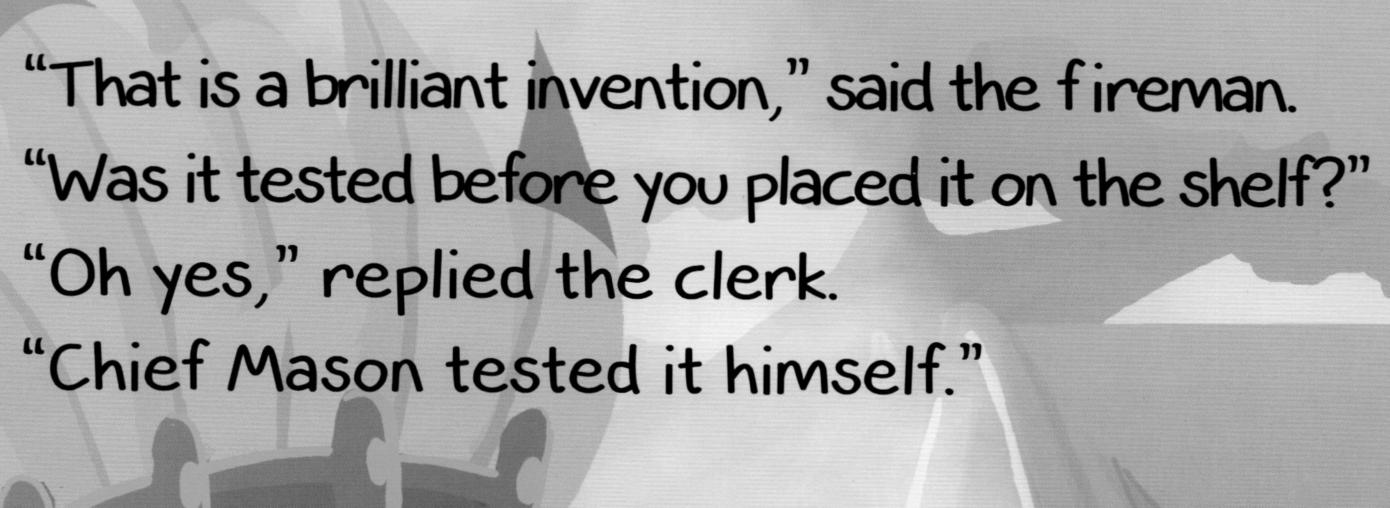

"That is a brilliant invention," said the fireman.
"Was it tested before you placed it on the shelf?"
"Oh yes," replied the clerk.
"Chief Mason tested it himself."

“He tested it by walking in a tent filled with smoke.
He walked around for 20 minutes and never even choked.”

The clerk added, “I will tell you a secret, but please don’t tell my boss.
If you want one of these, then I can spare you some of the cost.”

"Please, don't say a word and get me into a mess,
But you can buy this mask at Garrett's store and
pay one nickel less."

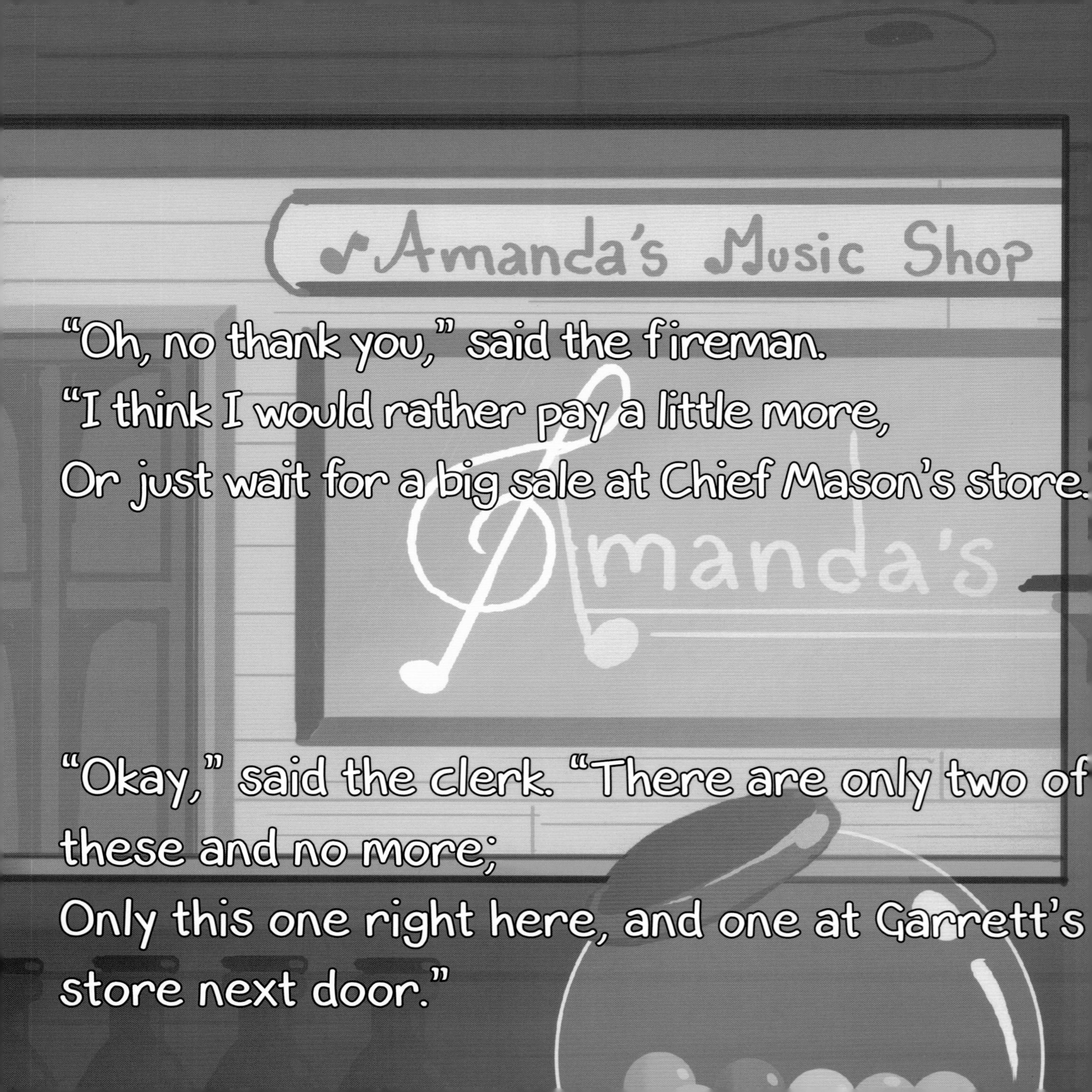

"Oh, no thank you," said the fireman.
"I think I would rather pay a little more,
Or just wait for a big sale at Chief Mason's store."

"Okay," said the clerk. "There are only two of these and no more;
Only this one right here, and one at Garrett's store next door."

Nicola
Nicola
Milk
Pasteur

Later that year, fire trucks were in motion.
Some workers were trapped in a tunnel explosion!

As smoke filled the tunnel it became too dangerous indeed.
The fireman who visited Chief Mason's store exclaimed, "I know just what we need!"

He remembered the invention and said,
"I know something that can help.
I will get Chief Mason to come down here
with that invention from his shelf."

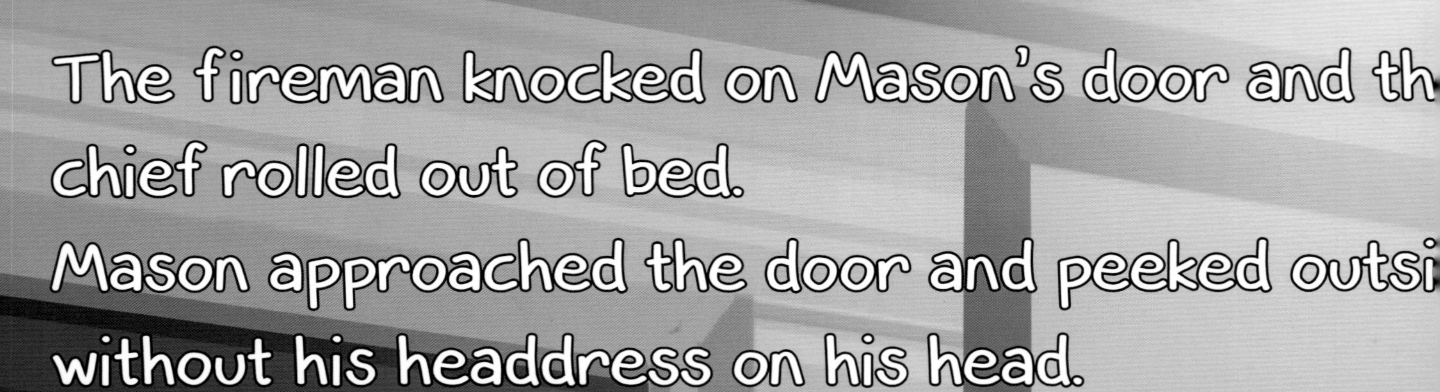

The fireman knocked on Mason's door and th
chief rolled out of bed.
Mason approached the door and peeked outsi
without his headdress on his head.

Mason heard the fireman yell,
“There is a tunnel filled with smoke!”
Mason placed his headdress on and knew he
was their only hope.

“Hurry Big Chief Mason, there is a
terrible disaster.
The smoke mask from your store can save
the workers’ lives much faster.”
Big Chief Mason answered, “I will meet you
there, but first I must wake my brother.”
Mason got one mask for himself and his
brother got the other.

Big Chief Mason and his brother ran straight into harm's way.
The people from the town held hands, hoping Mason could save the day.

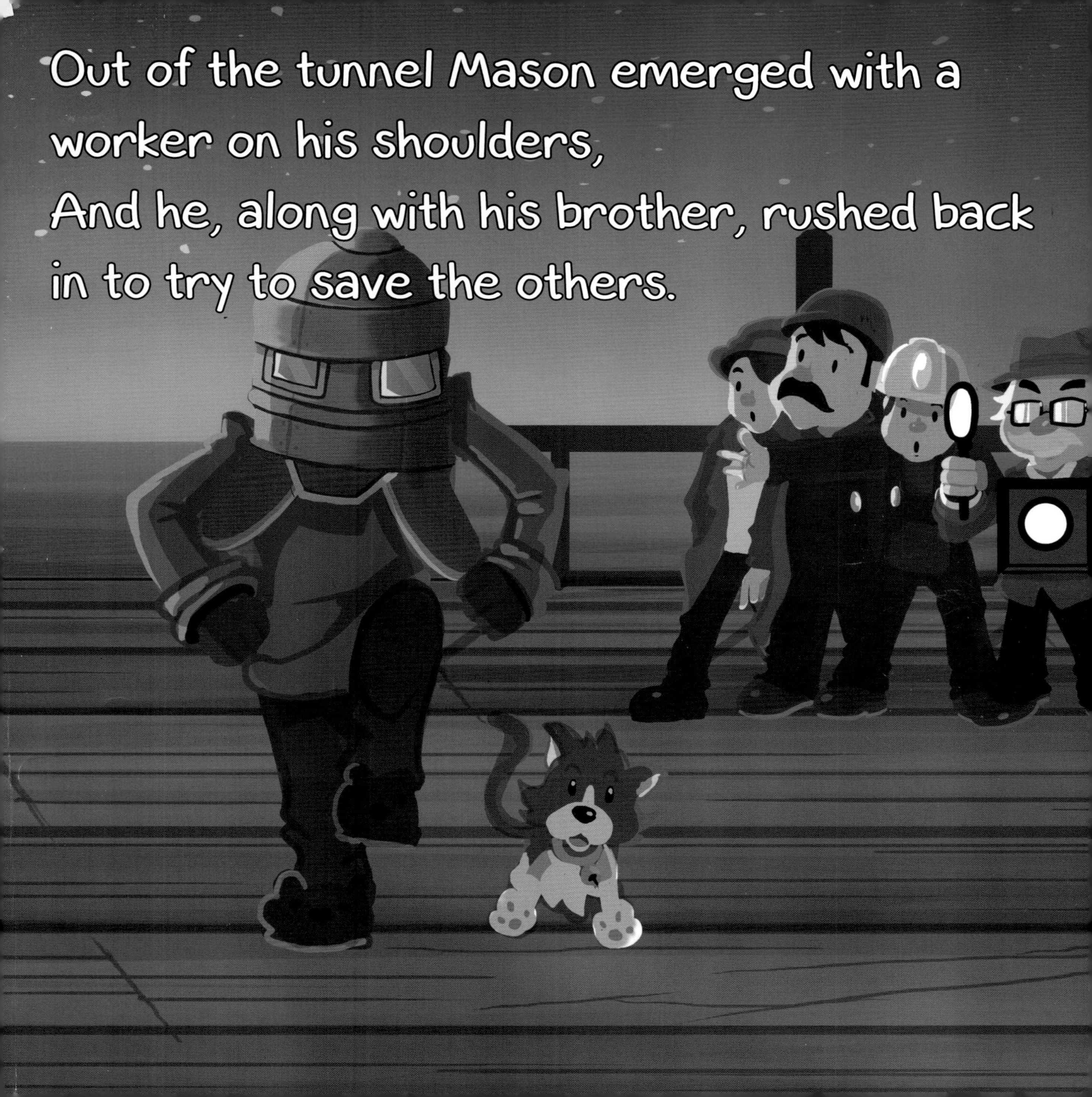

Out of the tunnel Mason emerged with a worker on his shoulders,
And he, along with his brother, rushed back in to try to save the others.

When they removed all of the workers,
a cheering crowd gathered all around,
As Mason took off his hood and shocked
the entire town.

The crowd stared at Garrett without a headdress covering his hair.
He did not wonder what they were thinking; in fact, he did not care.
See, Big Chief Mason was not who people thought he was at all.
He was just Garrett with a headdress,
No more, no less.

CLOSED

that day, Garrett did not dress up as Big Chief
anymore.
sed down his other place known as Big Chief
's store.

People came from miles around to see Garrett's great inventions.
His store filled up after the day his heroic act was mentioned.

The shelves at Garrett's store were
becoming quite a mess,
As people shopped at Garrett's
store and paid one nickel less.

Sometimes we hide who we are so that we can be accepted.
When Garrett took off his mask, the reaction was expected.
We may have straight, long, or curly hair; fine clothes, or hand me downs.
Maybe we speak a different language, or our skin is white or brown.

People may make fun of you, and treat other people wrong.
Remember that when Chief Mason took off his mask, he was the same lovable, inventive, heroic Garrett all along.

THE END

Garrett grew up to
be known as...

Garrett Morgan
1877–1963

Garrett Morgan was born in Cleveland, Ohio, on 4 March, 1877.
Garrett was an African-American inventor.
He invented hair straightening products, improved the traffic signal, and a safety hood that allowed a person to walk around a smoke filled room.

People say he invented a gas mask, but Garrett would not say it was a gas mask because many gases are heavier than air and his safety hood allowed air near the ground to be breathed in.

Garrett would have to hire actors to pretend they invented his products. He would pretend to be the inventor's helper and dressed up as Big Chief Mason, a member of the First Nations Walpole Tribe in Canada.

In 1916, Garrett, along with his brother and a couple volunteers, saved several workers trapped in a water intake tunnel below Lake Erie. Two previous attempts had failed. Garrett's safety hood was nationally known after this event, but local news outlets where the accident happened ignored his heroics.

In 1917, he was finally fully recognized as the hero he was, and he was presented with a diamond-covered medal.